Oliver Lodge

The Foundation of Dynamics

Elibron Classics
www.elibron.com

Elibron Classics series.

ISBN 0-543-69961-7 (paperback)
ISBN 0-543-69960-9 (hardcover)

This Elibron Classics Replica Edition is an unabridged facsimile of the edition published in 1893 in London.

From the Philosophical Magazine for July 1893.

THE

FOUNDATIONS OF DYNAMICS.

BY

OLIVER LODGE, F.R.S.,

PROFESSOR OF PHYSICS IN UNIVERSITY COLLEGE, LIVERPOOL.

From the PHILOSOPHICAL MAGAZINE for July 1893.

The Foundations of Dynamics. *By* OLIVER LODGE, *F.R.S.*, *Professor of Physics in University College, Liverpool**.

PART I.—*The Nature of Axioms.*

IT is a matter of congratulation with me that a critic who has devoted so much thought to Newton's laws of motion and similar fundamental doctrines should have begun a discussion of my papers on the subject of Energy; and I shall willingly consider his objections in order to see what modifications, if any, should be made in my original statements. But Dr. MacGregor's temporary attitude towards Physical problems is exhibited rather strikingly in a treatise on "The Fundamental Hypotheses of Abstract Dynamics," which he published as a Presidential Address to a section of the Royal Society of Canada (Transactions 1892). Hence, before replying to his criticisms on my writings, as made in your February issue, page 134, I should like to make a few general observations suggested by this other deliverance of his, so as to indicate what seem to me the rather different points of view from which we, or if not we some other writers, approach these fundamental doctrines of Mechanics and Physics. The difference in attitude may be briefly summarised thus:—Some philosophers seek to advance truth by detecting or inventing complications in what was apparently simple; whereas others aim at making simple statements concerning things which are apparently or really complicated. A generalization like this is

* Read before the Physical Society on May 12, 1893.

not one which will bear pressing into individual cases, but if it contains an element of truth it has reference to no personal detail, as it seems to me, but to a difference in type; and I sometimes think that most minds, except those few of the very highest order who are above classification, may be said to fall into, or at least to lean toward, one or other of these categories*. Each type of mind performs its service, and each type has its appropriate danger.

The detection of a real complication is a service to truth; the invention of a needless complexity is a disservice and temporary obstruction. The reduction of apparently complex facts to a simple statement in commonplace language is, I believe, a service; the over-simple and incomplete summary of what is really complex is not an equal service, but I do not perceive that it is likely to be any serious obstruction: it seems to me rather of the nature of a first approximation, which is often temporarily helpful.

When Ohm stated his law that current is proportional to E.M.F., he did not know that it was really true. It has turned out to be precisely true for copper and for sulphate of copper—the only substances for which it has been seriously tested; but even if it had not been so accurate, its statement was a service, since it enabled half a century to walk in the light instead of in the dark. There is no evidence that it is accurately true for every variety of solid and liquid conductor, but by this time it is the fashion to assume its truth in ordinary simple cases. And rightly so, as it seems to me; the burden of proof rests now with the enterprising experimenter who can detect a flaw in it. His evidence will be listened to, but till it is forthcoming vague doubts can be legitimately ignored.

Take another example:—The characteristic equation of gases in the simple form $pv=RT$ has done good service, though it turns out to be untrue for every actual substance. Without it, however, we should have been unnecessarily floundering in the dark. Even now it is more used in dealing with gases

* I see no reason in Dr. MacGregor's book on Dynamics for including him in the first category: it is his Presidential Address on the Laws of Motion that alone suggested it. I do not intend the classification as in any way offensive: I should think that Prof. Karl Pearson, for instance, would willingly enrol himself under the first head rather than under the second, judging by his 'Grammar of Science.' But very likely MacGregor has stated the laws of motion in their simplest conceivable form if attraction and repulsion across a distance are to be contemplated. That is the essential difference between us: he is willing to base Physics on action at a distance; I am not. From the action-at-a-distance point of view his statements are in many respects admirable, especially those near the conclusion of his essay. The remarks in the text are intended to have only a general and impersonal application.

than any other equation. The improved statement of Van der Waals, adding a term to p and subtracting one from v, was another distinct service, and enabled a mass of experimental evidence concerning the structure of liquids to be conveniently and simply summarised. In its turn, however, it has had to give place to more complex empirical statements, and the complete law has not yet appeared.

The examples I have chosen, one of a precise, the other of an approximate, simple statement, are not indeed of the nature of axioms; and it may be held that it would be unsatisfactory to base our axioms on such a tentative sort of footing.

And yet what other course is open? Truly axiomatic statements can only be effectively made concerning things of which the race has had a long course of experience,—things to which they have grown familiarly accustomed. If they can be actually proved, they are theorems, not axioms.

The setting forth of an axiom I regard as a kind of challenge, equivalent to the statement:—"Here is what seems to me to be a short summary of a universal truth; disprove it if you can. I cannot prove it, it is too simple and fundamental for proof, I can only adduce hundreds of instances where it holds. I have indeed critically examined a few special cases and never found it fail, but a single contrary instance will suffice to overthrow it; hence, though it be hard to prove, yet if not true its disproof should be easy: find that contrary instance if you can." If no disproof is forthcoming for a few generations, the axiom is likely to get accepted. Meanwhile its undeniable simplicity is a practical advantage, even though in the course of centuries a flaw or needful modification in its statement may be discovered.

This is the kind of basis on which such a law as that of the Conservation of Energy or the Conservation of Matter rests. That the perpetual motion is impossible, that matter is indestructible, that energy never diminishes in the act of transfer, all these must be regarded as generalizations based on a great series of experiments, some consciously directed to the upsetting of one or other of the laws if possible, some aimed at establishing them, but most of a non-contentious and collateral character. If we are challenged to produce direct evidence that in any given chemical reaction the mass of the reagents is unchanged, not only in the initial and final stages but at every stage of the process, the proof may be exceedingly difficult. Heat is liable to be developed which would interfere with delicate weighings, and the reaction challenged may be an explosive or otherwise inconvenient one. But we do not attempt the proof, we shift the burden on to the shoulders of the doubter and say to him, Disprove it if you can ; and so we practically say for all our axioms, and for all laws which are so simple and fundamental as to be hardly distinguishable from axioms.

Experiments are often made or adduced in support of a law as if they were part of its foundation : thus Newton tried experiments on impact before stating his third law, but the experiments did not really prove it with accuracy even for the particular case examined. All they could show was that there was nothing obviously wrong with it. He saw no reason for supposing it wrong, and so after consideration stated it as an axiom, to be hereafter challenged and found inaccurate if so the progress of experience turned out.

I should say that an axiom or fundamental physical law is a simple statement, suggested by familiar or easily ascertained facts, probable in itself, readily grasped, and not disproved or apparently liable to disproof throughout a long course of experience.

If a statement is capable of exact examination and verification, either by reasoning or by experiment, it is called a law, but not a fundamental law; *i. e.* it is no longer part of the foundation, it is supported on something else. If it has no support, except the absence of evidence against it, it is an axiom. Far be it from me to decry the use of experiments of verification. The necessity for them whenever feasible is conspicuous and universally admitted, and much ingenuity may be usefully spent upon them ; but I do say that in time a theory can become established by processes other than direct experimental verification; and in fact that really valid and flawless experimental verification is frequently an impossibility.

An instructive example of the legitimate strength of a theory, even when opposed by apparent facts, is contained in an article by Lord Rayleigh in the Philosophical Magazine for March 1889, "On the History of the Doctrine of Radiant Energy."

It appears that W. Herschel conceived the idea that the radiation which excites the sensation of light and the radiation which produces heat in black bodies are essentially different ; and this view, which was contrary to his original intuition, was based upon a crucial prismatic experiment, made for the purpose of ascertaining "whether the heat of the red rays is occasioned by the light of those rays" or not. A definite question, *answered by experiment in the negative.* On this Lord Rayleigh remarks:—

"I am disposed to think that it was this erroneous conclusion from experiment, more perhaps than preconceived views about caloric, that retarded progress in radiant heat for so many years. We are reminded of Darwin's saying that a bad observation is more mischievous than unsound theory. It would be interesting to inquire upon what grounds we now reject the plain answer which Herschel thought himself to have received from experiment. I do not recall a modern investigation in which the heat and light absorptions are proved to be equal for the various parts of the visible spectrum. Can it be that after all we have nothing but theory to oppose to Herschel's facts?"

Yes, that is all, and, as Lord Rayleigh well knows, it is amply enough. Whoever examines the facts again will do so not to substantiate our present theory but in the hope of upsetting it. Success is of course just conceivable, but, when it comes, there will be time enough to reopen the question. Lord Rayleigh's words may be distorted, and may even suggest false meanings to minds with a crooked turn in them; so may many of the apparent admissions about unprovableness in this paper of mine, but, whether it gives occasion to the enemy to blaspheme or not, it is true that a host of doctrines are believed because they form part of a consistent scheme rather than because of any seriously attempted, still less any really achieved, experimental proof. And to pull one of these neatly fitting blocks from its niche will demand the strength of more than one, of more than several, so-called crucial experiments. There comes a time indeed when the weight of experimental evidence suffices to uproot a portion of tightly fitting theory; but seldom, as I think, without some looseness or uneasiness being first detected, and never without a betrayal of rottenness at the root.

Why do Physicists deny that matter can be moved by mental power from a distance without physical mechanism? Why does modern science reject the whole of a certain class of miracles, in the teeth of an immense record of direct experimental evidence? Solely because these things do not fit in with such coherent views of the universe as they have at present been able to frame.

Why, again, do we accept a multitude of unverified statements, such as, that every portion of radiation, whether it be of light or of sound, is intrinsically energy, and must, if absorbed, result in heat; that every muscular contraction of an animal corresponds to the combustion of a portion of his food; that a given gas consists of particles of approximately specified size and weight travelling at a certain average speed; that a medium connects every pair of bodies which are perceived to exert force on each other; and so on? Solely because these things do fit into a coherent and self-consistent scheme of the universe.

Any scheme or doctrine *sufficiently* harmonious and consistent goes far thereby towards establishing itself as truth. So conspicuously is this the case, when one comes to reflect, that there are not wanting some who conjecture that by our thought we are, so to speak, *constructing*, or at least helping to construct, the cosmic scheme.

Some axioms the human race has now given up challenging, and by so abstaining has silently accepted as corresponding to the truth of things. Others it occasionally exercises its ingenuity in degrading or depreciating, not into untruths, but into special cases of a higher and super-sensuous generalization. Varieties of space are imagined, and mathematically treated, where more than one line can be drawn through a point parallel to a given line, where the shortest distance between two points is not straight, where the three angles of a triangle are not equal to half a revolution, where a closed surface is an incomplete boundary, and where more than three lines can be perpendicular to each other. These things are imagined, and for all I know they may in some occult fashion exist. To set bounds to the possibilities of the universe on the limited evidence of our few sense-organs would be absurd. But I say that any proof of their actual existence within our more developed ken rests with the experimentalist. As soon as facts are forthcoming which clearly and definitely are inexplicable on the basis of our present notions concerning space, I for one am willing to enlarge those notions and to contemplate provisionally whatever hypothesis suggests itself as most simple and plausible. Till then there is plenty of work for a physicist in interpreting, systematizing, and clarifying the facts of the universe, as it appeals to him through the agency of his ordinary three-dimensional senses, aided by his undimensional common-sense.

But I hold that with all these vague possibilities of ultimate development in front of him (not so vague but that they are in some sort conceivable, or at least tractable by reason), a natural philosopher need not confuse himself by endeavouring to complicate what is already transparently simple; nor will he be wise to attempt an over-laborious scrutiny of his fundamental axioms; for the more neatly and quietly he can lay his foundations the more time will he have for building the superstructure, and the more gorgeous he may hope to make it. By all means let him avoid a rotten or insecure element in his foundation. It must be as sound and strong as possible;

but his underground work need not be decorated with fanciful and laborious conceits, it may be as plain as it is substantial, and he may leave its edges rough in order to connection with structures as yet unbuilt and unimagined.

Of this plain and substantial character would I seek to keep the laws of motion. If a statement like the first law of motion cannot be made in simple and readily intelligible language, I should despair of Physics. In that case the Physics of the future could be little better than a barbarous jargon of technicalities. There are plenty of really difficult places where technical language and unfamiliar modes of thought are for the present essential. The developments of the superstructure erected during the present century are indeed now so stupendous that for myself I should be satisfied if, without appreciably adding to them, I could by consolidation and restatement remove the necessity for some of this artificiality, and so make their harmony and beauty more readily appreciated. But in order to do this a simple and unlaborious foundation is a necessity. I hope to try before long to display the bold outlines of the foundation already laid by men of surpassing genius, every unnecessary accretion being cleared away, and the whole simplified to the uttermost; and then, if the attempt be not too ambitious, I should wish to extend the process to some portion of the superstructure.

Part II.

The First and Third Laws of Motion.

So much for general preamble; now turning to Dr. MacGregor's address, we find that he objects to the Newtonian statement of the first law of motion on the same ground that Prof. Karl Pearson, Prof. Mach, Mr. Macaulay of King's College, Cambridge, and several others* have objected to it, on the ground, namely, that uniform motion is unintelligible or meaningless unless you specify its direction and velocity with reference to a set of axes. And directly you try to specify axes you get into difficulties, for, although a uniform translation is permissible to them, any rotation or any acceleration of the axes is fatal to a simple statement of the behaviour of a body acted on by no force. It is useless to say that the axes must be stationary, because one cannot define what that means; so the attempt is made to say that the axes must not rotate and must not be acted on by force; but this last condition is of no use unless they possess

* See for instance a correspondent in 'Nature,' vol. xxxvi. p. 366.

inertia; so the axes are sometimes supposed to be generated by particles of matter projected in different directions and subsequently free from force *; or else the axes are made of infinite mass so that no finite force may be able to affect them.

Now I hold that all such notions as axes of reference are artificial scaffolding, necessary for the numerical specification of a velocity, but not at all necessary for the apprehension of what is meant by a uniform velocity. It is in the specification of any absolute velocity that the difficulties cited about an origin and axes of reference legitimately occur. It is in fact impossible to specify the absolute velocity of anything, because we have literally no criterion of rest. We shall, I believe, hereafter find it convenient to postulate the Ether as a body absolutely at rest; but none of these physical or geometrical complications should enter into an axiomatic statement.

All that the first law asserts is that the motion of a body not acted on by force is uniform in magnitude and direction. There is no need to attempt the impossible and say what that magnitude and direction absolutely are. Whatever they are they remain constant. If asked to prove this statement, we should at once decline, and throw the burden of disproof on the doubter. This is what Maxwell does † when he says (virtually):—If the speed and direction of a freely moving body vary they must vary in some definite manner; very well, tell me in what manner they are varying. You cannot, unless you can show me absolutely fixed lines of reference.

The fact is that the conception of uniform motion is based upon a simple primary muscular sensation, or at any rate upon a succession of such sensations; everybody understands what it means, so far as it is possible to understand anything in this material universe, and the sense in which it is understood is amply sufficient as a basis for a physical superstructure. The first law is a true axiom, and its boldest and simplest form is not only the best, but is the only one that can with any justice be called axiomatic. How can one appeal to the experience of the human race with reference to coordinate axes of infinite or any other mass? How can we utilize as axes the trajectory of particles free from force, without tacitly assuming the first law continually? The whole attempt to complicate the statement of the first law of motion seems to me absurd.

* Thomson and Tait, vol. i. Part I. (1879), § 249. But although these writers do propose to use such axes to fix direction, or, better, the invariable plane of a rotating system, §§ 267, 245, they quite logically deduce these things from Newton's laws, and do not use them in the statement of those laws.

† Matter and Motion, art. xli. p. 36.

The well-known other objection, that a statement of the first law is unnecessary because it is only a special case of the second, rests on a different footing. Thomson and Tait have pointed out that it acts as a definition of equal intervals of time. Prof. MacGregor denies that the first law gives us any more useful definition of time than the second does; but seeing that every clock is an attempt at a uniform mover, and that the second law is concerned with the more complex notion of acceleration rather than with the simple idea of velocity, I do not imagine that he will seriously adhere to this view: and therefore, fully admitting the obvious fact that the first law is a special case of the second, I still hold that its separate statement is desirable, because it is so simple, and because it does afford a clear practical definition of the mode of measuring time. But were this its intended meaning it could have been expressed more straightforwardly. Newton probably considered it as a qualitative statement introductory to his second law: and as such it is entirely suitable.

None of the objectors to the first law have the slightest doubt of its truth,—that is what makes their contentions so practically futile; it would appear that it is too simple to please them; they seem to wish to complicate its statement so as to make it look more like the difficult things with which they are accustomed to deal. I feel convinced that many mathematicians mistrust a simple statement in English, and hardly conceive that such a language can really express an important law; their trained ability to deal with difficult conceptions leads them, as I think, astray.

I am not quite clear what the word logomachy means, but much of the discussion which has been bandied about concerning the statement of the first law seems to me to be rightly designated by some such opprobrious or perhaps complimentary epithet.

When Dr. MacGregor goes on to consider the third law, and to deny that it can properly be regarded as a deduction from the first, he is urging a very minor matter, if what I have said concerning axioms has any truth in it. Still the question has some interest. Whether it is deducible from the first law or not may be held to depend on how general the terms are in which that law has been stated. If it can be axiomatically asserted that the *centre of mass of a rigid system* moves uniformly until an external force acts on the system, and also that the system does not begin to spin, then the third law is established. For since zero acceleration means zero force, it follows that all the internal forces add up to zero, and have no moment; and since the system can be dissected bit by bit without ceasing to be a system within the scope of the first law, it follows that no stress can contain an unbalanced force or couple.

If the law be doubted for the case of a pair of bodies attracting each other from a distance, Newton says* (virtually):—Jam the bodies apart with a rigid obstacle, then you have reduced their action to contact action; and since you have a balanced stress at each point of contact, and likewise between the ends of the introduced obstacle, it follows that the attracting forces of the distant bodies are also balanced.

Now Dr. MacGregor's objections are (1) that Newton's proof only aims at extending the law from contact action to actions across a distance, while for contact action he is content to assume it as an axiom or to verify it by experiment; and (2) that the proof breaks down for a particle or single body which cannot be analysed into parts.

It is quite possible that Newton thought it best to state his third law as an axiom; because the fact that the centre of mass of a complex system of bodies obeys the first law is hardly an experience that can be confidently appealed to, even though all the bodies are in contact. That fact and the third law are intimately *connected*, but whichever is the simpler had better be stated as the axiom, and the other be made a deduction from it. Thus the centre of mass statement follows from the third law, and so very likely Newton preferred to arrange it. It is frequently difficult to know which of two very simple statements is the more axiomatic; and methods of proof are notoriously susceptible of considerable variation. The important thing is to notice the link or tie between two facts, to show that they mutually strengthen each other, not to pretend that one is beneath the other and supports it.

Attempts to build even so simple a structure as geometry in the form of a single column, stone upon stone, have been found artificial and in the long run impracticable. An enlarged basis of direct appeal to experience is not only necessary but desirable, and all fundamental matters should be kept low down, as nearly in contact with first principles as possible.

The so-called deduction of the third law from the first or second is important as a clear and strong cross-connexion between the two things, and need not be considered as a rigorous proof. It is rigorous enough if the premisses are granted, but if not, then there is a certain outstanding axiomatic or unprovable character to be shared between them; but this outstanding portion is, by reason of the cross-connexion,

* *Principia, Scholium* to *Axiomata.*

so slight as to remove all difficulty from what otherwise would seem, and indeed, strange to say, to many still does seem, an exceedingly tough morsel to swallow.

[I mean that one constantly finds examination candidates, and even Engineers, when catechized about a horse pulling a cart, though they may, some of them, politely admit that approximately the pull of the horse and the pull of the cart are equal (constituting the stress in the trace whose inertia we may agree to neglect), yet nevertheless assert that in reality the pull of the horse must be *the least little bit* bigger than the pull back of the trace, else the thing could not start. The fact is that the universal truth of the third law is *not* axiomatic, or at least is not obvious*, and hence its deduction from the other two laws is really a useful deduction.]

I do not see any point in Dr. MacGregor's second objection that the proof is inapplicable to a body without parts. For if such a body anywhere exists, plainly its parts cannot act on each other, and so there are no actions or reactions in such a body worth troubling about.

The concluding portion of Prof. MacGregor's address has to do with the Conservation of Energy and the question of how far it can be deduced from the third law. But all these questions respecting energy we must fight out at greater length. Dr. MacGregor well knows that from the third law and the denial of action at distance together I claim to have deduced a law of conservation simpler and more precise than the ordinary law; but his objection to this† is that though such a law may come to be accepted as sufficient in the future, when the universality of contact-action is fully recognized, it is inadequate for the present, when action at a distance still holds a portion of the field; by which I suppose he really only means that many mathematical methods of treatment are based at present on action-at-a-distance modes of expression. I have no fault to find with any convenient mode of attacking specific problems; it is permissible to everyone to use the language of distance-action for practical purposes; but when it comes to formulating fundamental laws I have no ambition to legislate for such cases until they can be shown actually to occur. I am open to experimental proof of their existence, but to none other. It is premature to legislate for them. If any action other than contact-action exists, we had better know more about it before formulating its laws. If the will-power of a "medium" for instance can really move a chair without

* See for instance 'Nature,' vol. xxxvii. p. 558; and see all recent volumes of 'The Engineer' *passim*, especially about 1885 and 1891.

† Expressed elsewhere, viz. Phil. Mag. February 1893.

any kind of contact or physical connexion, it will be wise to look carefully for the seat of the other component of the stress, if there is one, and for the source of the energy concerned, but I myself should feel extremely hazy as to their probable locality.

When action at a distance does present itself in Nature (and if it ever does it is clearly going to be in connexion with the operations of Life), it will be very well to overhaul our axioms to see if they require modification. Till then I propose to state them in terms of the facts we know. This I will attempt in another Part.

Part III.

The Conservation of Energy and Universal Contact-action.

The ground is now clear, I think, for a reply to Professor MacGregor's criticism, as made in the Phil. Mag. (vol. xxxv. p. 134) for February 1893; and incidentally I may hope to answer or at least discuss the matter with some other critics, notably Mr. Heaviside in his paper in the Phil. Trans. 1892.

The first objection is that in my definition of energy I assume the ordinary law of conservation, because I say, in an early paper (Phil. Mag. Oct. 1879, viii. p. 278), "Whenever work is done upon a body, an effect is produced in it which is found to increase the working-power of that body (by an amount not greater than the work done); hence this effect is called energy, and it is measured by the quantity of work done in producing it." "The words 'is found,'" says Dr. MacGregor, "indicate an appeal to experience." Most true, so they do. My position is this:—Before making any definition it is desirable and only civil to show the reasonableness of it. To thrust a statement out without preamble or explanation, under cover of the contention that being only a definition or an enunciation one is at liberty to define or enunciate as one pleases, is I fear a thing frequently done, but it is barely polite, and it is apt to excite either resentment or else undue and slavish submission.

It is from no lack of love for Cambridge and the great men she has nurtured that I venture to hold that the typically Cambridge plan of text-book is liable to err in this direction. An attitude of blind faith and mere assimilation is required of a student for many of the earlier chapters, sometimes for the whole of a book. The life, the interest, of the subject to be treated, an exhibition of the reasonableness of the adopted mode of treating it, are all neglected, and a ghastly skeleton is presented to you bone by bone; which truly is an admirable structure when subsequently clothed in flesh by men of sufficient genius, but which is liable to excite repulsion in a thinking student who possesses some physical instincts but is devoid of the artificial galvanic stimulus called Examination*.

It may be that I myself err in an opposite direction, being so interested in the muscles and the clothing that I forget here and there a bone or two. If it be so it is a grievous fault, and grievously shall I have to answer for it. It is not a fault that I ever attempt to justify. Flabby and boneless science is no science at all.

In defining Energy, then, I first appeal to experience that something in the popular sense *energetic* is often plainly produced in a body when work is done upon it; *e. g.* when a bow is stretched or a stone flung; and I proceed to say that that something is always produced, even when not obvious, and that it has been called Energy because it frequently confers upon the body possessing it the power of itself doing work. I know well enough that the common definition runs, Energy =power of doing work, but there is a difficulty about this definition: plenty of energy has no power of doing work, or at least no power that we can get hold of. Therefore I prefer to give a name to the result of work done, whether it be obviously energetic or not, and to justify the name energy by appealing to the many cases where its possession does confer a power of doing at any rate some work; but in the definition I make no statement that energy must necessarily be able to do work, or must necessarily continue constant in quantity, because such statements, so far as they are true, are part of the law which is being led up to, not part of the definition.

My definition of "energy" stands on all fours with the customary definition of the potential function. It is a name for the line-integral of a force, considered as a quantity that can be stored. The line-integral of a force in action is *work*, the result of it is *energy*. Or, otherwise:—The scalar product of force and velocity is *activity* (sometimes called power); the result of activity lasting a finite time is energy.

Some conception of what energy is like can be gained by appeals to familiar experience; and the fact that it can be stored more or less completely, and can ultimately reproduce an activity similar to that which generated it, can also be illustrated in a selection of cases. The law of energy asserts, what cannot be so readily demonstrated, that in *all* cases it can be stored without loss, except in so far as it leaks away, or until it is discharged, in some kind of equivalent activity; and that even when thus dispersed or transferred or lost to

* Valuable enough, however, in its proper place. I am not joining in the outcry against examinations. They are a most useful and much needed stimulus to right intellectual conduct, and partially replace the old belief in purgatory.

the body in which it was generated, it still exists in other bodies in undiminished quantity, and so is capable of transference and retransference, by means of activity, and activity alone, for ever.

But how is such a wide-sweeping statement to be justified? Not by any hasty appeals to experience, not even by the laborious researches of a Joule. Such experiments can indeed examine the obviously weak places and can show that there is no manifest flaw there, that all the apparent flaws disappear on closer inspection, and, so far as can be seen, do not exist. The law itself as a universal generalization must be itself axiomatic, or else must be deduced from some other and simpler axioms.

I hold that it can be deduced from Newton's third law and from the denial of action at a distance, thus:—Bodies can only act on one another while in contact, hence if they move they must move over the same distance*; but their action consists of a pair of equal opposite forces; therefore the works they do, or their activities, are equal and opposite; therefore, by definition (p. 13), whatever energy the one loses the other gains. In other words: in all cases of activity, energy is simply transferred from one body to another, without alteration in quantity.

I claim also that the law of conservation thus established is more precise than the ordinary law (which I confess always seems to me rather vague, especially when such absurdities as "possible" and "actual" energy are put into its statement,—a denier of the law could use no more deadly word than "possible"); more precise and definite, I say, because it is the law not only of *conservation* but of *identity*. I believe that questions arising from this law of the identity of energy—a study of the paths by which any given now-existing bit of energy has reached its present locality, and of all that has happened to it in the past—may prove in the future to be as fruitful a region of enquiry as are studies in the history of any given piece of matter, say the earth or the sun, or, to step forward a little, the history of an individual mind, when we realise some day that this, too, has a continuous, and perhaps traceable, existence.

However, Prof. MacGregor objects (p. 135) that my deduction of the conservation of energy proves it during *transfer*

* Motion in the line of stress is here directly contemplated. As has been already several times noticed, the possible *slipping* of bodies in contact is not apparently allowed for, but in so far as they are smooth their slip is ineffective, and if they are rough there is a tangential stress to be taken into account as well as a normal pressure; the friction-stress is intermittent instead of steady, and molecular instead of molar, but otherwise the above statement applies without complication, so far as it ever applies to the immediate action of ordinary material bodies.

merely, and leaves untouched the question of what happens during simple *storage*.

That is true, and I do not think that I had noticed the objection before. My deduction proves that in all cases of activity energy passes on without loss ; it does not prove that without any activity energy may not leak away in some silent unobtrusive fashion, leaving no trace, but simply vanishing.

To meet this objection, an appeal may be made to the definition; it may be called on to assert that anything that can thus disappear without force or motion is not energy, according to its terms; but I doubt whether such an appeal is perfectly cogent, though it has some, perhaps much, plausibility. I think that the non-disappearance of energy in this occult manner had perhaps better be regarded as an axiom, a statement which may be believed until some clear experimental disproof is forthcoming: no such disproof being meanwhile in the least expected.

Briefly summarised, the matter stands thus :—

(1) If energy can only be got rid of by activity, and (2) if that activity never affects its quantity, the law of conservation is completely stated. My proof covers (2), but not (1). I may therefore agree with what I suppose Prof. MacGregor to mean, that the portion (1) must be left as an axiom based on experience.

Prof. MacGregor goes on to say (p. 137) that my law of conservation during mutual action has no deeper meaning than the conservation of their joint momentum, and is quite consistent with the non-conservation of their working-power.

To this I reply, first, that I mistrust the vague term "working-power," it is apt to mean whatever may be convenient ; from one point of view a given amount of energy may have an infinite "working-power," since it can do work at every transfer without itself diminishing; while from another point of view it is rather bold to maintain the conservation of working-power in face of the doctrine of the dissipation of energy. And secondly I reply, that the conservation of *momentum* rests on the equality of the forces exerted by two mutually operative bodies, combined with the obvious equality of the *durations* or times of action of these forces. Whereas the conservation of *energy* rests on the equality of the forces, combined with the altogether less obvious fact of the equality of *velocities*, or *distances traversed*, by each of two operative bodies. Anyone in his senses who believes in action at a distance would deny that the velocities of two directly acting bodies are necessarily equal (no other bodies being assumed present, so as to simplify the problem); and in so doing he practically denies the conservation of energy.

Energy is only really conserved under conditions of universal contact-action.

Of course the believer in action at a distance is not at a loss: in order to retain his fiction he has invented another unreality, which he calls possible energy. He says (truly enough for many purposes) that when a stone is raised from the earth a great deal more work is done on the stone than on the earth; and hence that, although at every instant the stone and earth have equal opposite momenta, the energy is nearly all possessed by the raised stone. But since the energy of an inert mass is by no means apparent, since it only has the power of gaining actual energy when let drop, its energy when inert and merely elevated is called possible or potential. Or sometimes, rather more accurately but much more vaguely, energy is said to belong to the *configuration* of earth and stone *.

Now this idea of potential energy is convenient as a mathematical expression, and used in its proper signification it is an essential reality; but used in the sense above quoted, a very common mode of using it, it is a mere receptacle for stowing away any portion of energy which it is not convenient for the moment to attend to; and I defy anyone to realize it as a thing.

Yet I myself constantly employ the term potential or static energy, and assert that every activity not only transfers energy from one body to another, but also transforms it from kinetic to potential, or *vice versâ*.

Yes, but I also assert that the transformation can only accompany transference, and that transference cannot occur without transformation. Whereas, on the ordinary view, the energy of the raised stone is supposed to gradually transform itself into kinetic as the stone drops, but to remain in the stone all the time.

I say that the energy was no more in the stone when merely elevated than it is in a strung arrow or the bullet of a cartridge. The energy is in the bow, or in the powder, and is rightly said to be in a static, or strain, or potential form. It can transfer itself to the projectile, and simultaneously transform itself into kinetic, at the pull of a trigger †.

* *Cf.* Phil Mag. June 1881, p. 532, and June 1885, xix. p. 484.

† On page 141, Prof. MacGregor makes a statement in which I entirely fail to catch his meaning. It runs as follows:—"When Prof. Lodge states that 'a bullet fired upwards gradually transfers its undissipated energy to the gravitation medium, transforming it at the same time into potential,' he seems to me to assume that the bullet is rigid and that the medium is without inertia." Surely my critic does not consider that either plasticity in a projectile or inertia in a gravitating medium is essentially involved in determining the height attained by a body thrown up.

It is easy to express oneself so as to be understood to object to the whole idea of potential energy. It is not to the use of the term that I object, nor even to its misuse, so long as it be done in the temporary interest of some specific problem. It is by no means necessary always to attend to everything: acting mechanism is frequently, with convenience and brevity, ignored, especially when one is really ignorant of its dynamical nature; but in laying foundations we should not make these omissions and slurrings. It is the erroneous localization of energy in a fundamental or theoretical treatment of the subject that I deprecate, and I believe that everyone will agree that in all such treatment convenient fictions are better avoided.

There certainly exists potential energy in the case of a raised stone, but it does not belong to the stone; it belongs to the medium, whatever it is, which is exerting force on the stone and on the earth, and pressing the two lumps of matter together. It is properly called static, as distinguished from kinetic, because it is the energy of force, not the energy of motion *. As I have frequently pointed out, the two forms of energy correspond to the two factors of the product work or activity. Both factors are necessary for the actual performance of work. Until both factors are present, the energy is merely stored. As soon as the missing factor is supplied, it is *transferred* from the body acting to the body acted upon, *e. g.*, from the gravitation medium to the stone.

But it will be said, suppose the stone is not allowed to drop freely, but is used as a clock-weight; it is doing work all the time it slowly falls, hence when raised it must have had energy. But in this case I should deny that the stone is doing anything active; it is a necessary concomitant, it is a link of communication, it is not really itself doing work. Its presence indeed and that of the earth are necessary to the existence of the stress in the ether [permitting that or some other equivalent hypothesis for the moment]; but it is the ether stress which is doing the work and driving the wheels of the clock. A

* I do not mean to assert or deny anything concerning the nature of gravity. It may be that all energy is *ultimately* of the nature of motion, as we know Lord Kelvin has taught us may be the case with elastic stress for instance; but *force* is the proximate mode of action of a coiled spring, and so it may properly be described as static energy, whatever its ultimate nature. The reasons for affixing to static energy the epithet potential are mainly mathematical and historical. There is much convenience in thus linking it on to the potential function, and there is no particular inconvenience unless the adjective be misunderstood in the sense of "possible." It is indeed possible *work* or possible *activity*, but so is every kind of available energy; it is not possible *energy*, for that term is equivalent to the assertion that not yet is it energy at all.

clock-weight has somewhat the same function as the string of a bow; it was the convenient means of generating the potential energy, and it is the means of communicating it to the thing to be driven, but it is not itself energetic, any more than the chain and ratchet-wheel which helped to wind it up were energetic. Or, to take another example, it is like the piston of an air-gun: without it the gas would not have been compressed, and it might be (though it is not usually) employed to let the pressure reversibly down again, but it is not itself an active agent either in the charge or the discharge. It is like a Holtz machine or a dynamo: needful to the charging of Leyden jars or storage-cells, but not in itself doing the work, either when used for charging or, when running as a motor, for discharging.

I fear I am explaining elementary matters at some length; but, although elementary in one sense, they are not quite easy to seize by the right end; and the fact that they are so elementary tends actually to retard their apprehension, for unless people have patience to think them out they will not grasp what is meant. The difficulty in all these matters is that everybody thinks he understands them already, and is quite satisfied with his own prehistoric way of regarding them. Now I want to say that I have thought these things out with some care and labour; and I believe that although no doubt everybody does understand them sufficiently for practical purposes, yet, if anyone has faith and patience enough to consent to reconsider them from my point of view (assuming, of course, as in most cases I safely may, that it is not his already), he will sooner or later realize the advantage of it. This belief may be presumptuous, but if so I am willing to presume to that extent.

With the substance of the following quotation from Professor MacGregor, on p. 138, February Phil. Mag., I entirely agree.

"If there be actions in nature which are not actions at constant distance" [*i. e.* practically contact-actions] "Prof. Lodge's law is not applicable to them, while the ordinary law is. Even if it be admitted that all actions in nature are contact-actions, there are many groups of phenomena which, in the present state of our knowledge of them, cannot be investigated on the hypothesis of contact-action. The early stages of their investigation must be conducted by the aid of the fiction of action at a distance; and in such stages Prof. Lodge's law is not applicable, while the ordinary law is. Hence Prof. Lodge's law is not so general in its applicability as the ordinary law."

I might prefer to express it differently: I would say that it is no discredit to the true law not to lend itself to fictions; but I would also say, and freely admit, that it is often permissible for practical purposes to work in a fictitious or incomplete manner (as, for instance, when dealing with the diurnal motion of the sun or the libration of the moon), ignoring communicating mechanism in cases where one is ignorant of or need not attend to its nature, and in fact proceeding as the great physicists have constantly proceeded when engaged on a practical quest.

Never have I imagined that the practice of men of genius is other than the practice most suited to their immediate end. But when it comes to the philosophy and essence of the matter, I do think that there is some fear lest a constant habit of ignoring unknown essentials for practical and temporary convenience should lead these great mathematicians into forgetting that they have ignored anything, should in the long run delude them into treating their fictions as if they corresponded to the reality of things; and it is permissible for me or others, as lookers-on seeing the structure from another point of view, unobtrusively to point out to the master builders, in their moments of leisure, the fact that while they have been engaged in the upper stories the foundations have almost automatically expanded, that the available basis of axiomatic reality has become broader and simpler than could be perceived when they started on their work, and that now some portion of their temporary scaffolding and underpinning may without danger, and with advantage to the general appreciation of the fabric, be discreetly and quietly removed.

Part IV.

The Dissipation of Energy, the Nature of Potential Energy, and the Second Law of Thermodynamics.

Passing on to Prof. MacGregor's fourth objection (p. 140) that I have not proved that transformation and transference necessarily go together, I hope to be able to meet him more nearly than has been possible to me on most of the other counts of his indictment.

My proposition was that the change of form is always from kinetic to potential or *vice versâ*. But it is not necessary to attach the same importance to this law as an ultimate fact, because it is so extremely probable that most or all cases of potential energy will ultimately be resolved into essential motion. Nevertheless it is for the present a convenient statement of truth, having the same sort of validity that the second law of Thermodynamics possesses.

The following is my demonstration of the law of transference and transformation: being indeed the detailed statement of the essential phenomena accompanying transfer of energy, on the hypothesis of contact-action.

A body or other medium exerting force at any point, and there moving in the sense of the force it is exerting, is in a state of activity: it is passing energy on ahead. If its speed continues constant, the force just spoken of cannot be the sole force, the resultant force on it must be zero; in that case it is a mere transmitter, not itself active, only passing on what it receives. But if it is itself active,* i. e. *parting with its own energy, then its speed at the acting point (the place of application of the force) must either decrease or increase.*

If its speed decreases, it must be parting with kinetic energy; if its speed increases, it is parting with potential energy.

Contrariwise :—

A body or other medium exerting force and being moved in the sense opposed to that force is receiving energy.

If it is not merely passing it on, in which case the resultant force acting on it is zero, its velocity must vary.

If the velocity increases it is gaining kinetic energy; if its velocity decreases it is gaining potential.

Now the two bodies or things here spoken of are necessarily existent in every case of activity; one is the agent the other the patient, one the emitter the other the receiver, one the acter the other the reacter, one the driver the other the driven; and they are in contact while the activity lasts.

Being in contact, their velocities along the line of stress must increase or decrease together.

If their common velocity is decreasing, then the driver loses kinetic and the driven gains potential.

If their common velocity is increasing, the driver loses potential, the driven gains kinetic.

This is my proof of the necessary concomitance of transfer and transformation, and of the alternation of transformations. But it will be well to illustrate the matter further.

Let me first explain how I define potential or static energy as used in the proposition criticised. I mean by it simply the energy of a body under stress; an elastic body which is exerting force may always be said to have potential energy, notwithstanding that when the force comes to be analysed it

* Meaning by "a body" a thing of constant mass, an identical lump of matter. A thing of variable mass, like a rain-drop or a railway-train or a layer of moving gas, is gaining or losing matter as well as energy; and cannot be regarded as one simple *body*.

may be perceived to result from some kind of motion. Thus a reservoir of compressed air is a store of potential energy, though it is well known that all the pressure is due to the motion of unstressed particles, and is really just as much kinetic as any of the energy of planets and comets. A powder-magazine is a store of potential energy, though quite possibly the forces awaiting a trigger to liberate them into activity are due to atomic or etherial motions; and the same may be said of a charged Leyden jar and of a bent bow.

Briefly the idea of potential energy corresponds to the idea of elastic force when the cause of that force is not sought. The idea of kinetic energy corresponds to the idea of sustained motion. Each corresponds to one of the factors in the product "activity," or **Fv**.

In order that a body may possess energy it must be *capable* of exerting force and also of moving; but it need be *doing* only one of these things, and so long as it merely stores energy it *must* be doing only one of them. Both factors must concur before it can be active, both factors must be possible before it can be said to possess energy. The only question is, which factor does it possess meanwhile? If it is exerting force, but stationary, then its energy is potential or static; if it is moving quite freely, then its energy is kinetic.

So long as the other factor is absent no activity manifests itself, no work is done, and the energy is merely stored. Directly the other factor is supplied transference begins, energy leaves the body storing it, and passes to the body supplying the other factor.

Now to make a *moving* body do work, a resisting obstacle must be supplied, and this second body thus exerting force receives a due proportion of the energy; and it receives it by reason of the force it exerts, that is, it receives static or potential energy from a body which possessed kinetic.

To make a *strained* body do work, motion must be permitted; the thing which is moved gains energy, and it gains it in the kinetic form from a body possessing potential.

Of course in both cases the body immediately receiving the energy need not retain it, but may rapidly or instantaneously pass it on to something else—in which case it resumes its first form; by simple alternation of transformations.

Also it is not essential that the potential energy of the particles of a spring shall be communicated to any *foreign* body: if it is transferred to the spring as a whole, considered as a mass possessing inertia, that is sufficient; in other words, it may be transferred from one set of particles to another set of particles, all forming part of what we usually speak of as a single body*.

To take one of the examples of Prof. MacGregor, viz. a loaded air-gun with its muzzle plugged, so that its bullet or wad is not allowed to be shot. The compressed air has potential energy; on its release its energy is transferred to the moving wad, which instantaneously hands it on to the air near the muzzle, compressing it, and thus retransforming itself into the potential form.

If it be objected that the energy never at any instant all existed in the kinetic form, the answer is no, but it necessarily passed through that form; it could only be by the intervention of motion that the energy of the air at one end was transferred to the air at the other end. The wad itself is unnecessary and may be dematerialized; the only thing moving may be the gas atoms themselves, but the atoms which are rushing are not the same as those under strain; the moving atoms are precisely those which have escaped from strain. As everybody knows, where the rush of gas is greatest in a constricted pipe there the pressure is least. Air rushing from one reservoir into another illustrates the law precisely, though I choose it as a case at first sight favourable to Prof. MacGregor's contention; in the narrow part of the pipe air is flowing against a higher pressure in front of it, it is compressing the air ahead of it by reason of its own momentum; the energy of each molecule is availably kinetic to precisely the extent to which it ceases to be potential, and any residual potential energy which does not become momentarily kinetic, and kinetic in the available sense of a windy rush, has not been transferred at all, but remains simply stored, as it would have been had nothing been released and no activity been manifested.

If the two reservoirs (or two Leyden jars) are for instance equal in capacity, and if one had been originally empty, then one quarter the original energy remains still potential in the originally full vessel, another quarter has been transferred by the rush of gas into the originally empty vessel, and two quarters have been dissipated, so to speak, converted into heat by friction. If there is no friction, then this half of the original energy remains alternating for ever from the kinetic to the potential form, like the bob of a pendulum, and transferring itself at every half swing from the one vessel to the other.

Lest this last statement should permit misconception, it may be necessary to state explicitly that the energy is only hazily attributed to the vessel—the energy belongs of course

* This sentence is true in one sense, but it is not a final or complete statement, and in a later paper a modification will be introduced.

to the gas molecules—and, further, that the molecules which are in motion are not the same as those which are under strain. The energy, that is the portion of energy concerned with the particular amount of activity contemplated, is continuously leaving the molecules recovering from strain and transferring itself to those in accelerated motion, and as continuously leaving those in retarded motion and transferring itself to those acquiring strain*.

Further to avoid misconception it is necessary to admit distinctly that in speaking of the pressure-energy of a mass of gas we are subjecting ourselves to an unnecessary though convenient limitation. It is chosen as an example just *because* so much is known about it; it would have been easy to choose elastic solid examples, where the elastic stress is not so readily analysable, and where the idea of potential energy and the statistical grouping of a mass of molecules was not optional but necessary, in our present state of ignorance. I know well enough that the pressure of gas in a reservoir is really due to the motion of the particles, and I am willing freely to contemplate the case of Davy's "repulsive motion," no real contact or elastic impact between particles ever occurring, but only a rapid swing or asymptotic orbit round their common centre of gravity; but this rapid centripetal swing of particles getting within each other's molecular range is at present essentially an unknown process, and has to be relegated to the sphere of potential energy quite as much as in the not perhaps really dissimilar cases of elastic impact and gravitation.

Whenever an atom after collision retains its energy intact, there is no need to say that there has been either transference or transformation; mere retention or storage is sufficient for practical purposes (though if we attend to the details of a collision we find there really is always a double transference and a double transformation †); and if this is typical of what

* See, however, the footnote on preceding page.

† *E.g.*, suppose a quick molecule A strikes or otherwise collides with a slow but similar molecule B. The mechanism of the collision, being unknown, is unimportant, and is most conveniently thought of in terms of elasticity. A and B undergo distortion as they approach, and a certain portion of A's energy is communicated to whatever medium it is which keeps them asunder. This medium by its recoil then drives them apart, sending off B with the energy which A originally had, and transferring to A the small portion received from B. [This matter is discussed at greater length in Appendix 2.] Any part of the energy which was not transformed is also not transferred. Everything transferred from A to B necessarily underwent rapidly a double transformation. When other occurrences are analysed it will be found that the above is typical of what constantly happens. Yet it is not always necessary to attend to all details, and in the text above I speak as if A and B passed through each other, each retaining its original energy.

happens in every atomic encounter, whether it be with the walls of the vessel or with another free molecule, then, so long as the air remains compressed, there is merely a storage of energy, each molecule possessing and retaining its own share. But directly an opening is made, or a communication established with an empty vessel, it is equivalent to the removal of a portion of wall, and some of the atoms, with their energy, pass along without obstruction into a new region of space.

In one sense each may still be said to retain its original energy; but in order to regard it thus we must be capable of dealing with the atoms individually. If the atoms escaping from confinement are lost in the vastness of space, their energy is unavailable, and their pressure, on which entirely depends the potential energy of a compressed mass of gas, is *nil*. If they escape into another equal vessel, then only the first individuals can enter freely without obstruction, all the others experience encounters on the way, as if the hole in the wall were gradually being restored; and though at the end of the operation, when everything has finally settled down, each atom may be said to possess its original energy, yet the pressure, on which depends the potential energy of the compressed gas, has been halved, and the other half is unavailable or has been dissipated.

Plainly the idea of potential energy belongs to the temporary order of ideas, to which the dissipation of energy and the second law of thermodynamics belong, and is appropriate to the present period when we have not yet learnt how to deal with molecules individually.

The kinetic energy of a set of gas molecules is only really available when it is combined with momentum or angular momentum, *i. e.* when the motion of a majority of molecules has the same sign, when there are more plus terms than minus in either their translational or their angular velocity; in other words, when there is a wind or cyclone *. A stream of particles can be utilised, as by a windmill or a Pelton water-wheel, and that is an example of available kinetic energy. But usually the energy of fluids can only be

* It may be worth while to point out that the reason why momentum is necessarily preserved in cases of impact, while translatory energy is usually not conserved, is because momentum depends on the first power of velocity, and therefore is unconcerned with vibrations; while every kind of vibration, whether it be of sound or of heat, enters into the sum total of a thing depending on an even power of velocity, like energy, whose translational value is therefore to that extent diminished by the production of vibratory disturbances.

utilised in the potential form; as in a hydraulic-press or steam-boiler.

To say that all potential energy will turn out to be really kinetic may be true enough, but it is not at present a specially helpful truth, for even in cases (such as compressed air) when we actually *know* it to be kinetic, we are bound for all practical purposes to treat it as potential, since the available working-power of a spring, or of compressed air, of gunpowder, of Leyden jars, and of many other things, depends on their potential energy alone; and such other energy as they may possess is conveniently ignored, as not concerned in any practical transformations or activities which we can bring about: it remains as an untransferable, and therefore inactive or useless residuum.

This is the real meaning of available energy, it is the portion concerned in transferences, the portion which can be transformed, *the portion which is able to transfer itself.* Potential energy, as commonly spoken of, is always of this character. An unwound watch-spring has lost its potential energy, it retains a quantity of untransferable heat-energy. A hot body cooling is in the same case; so long as it is hot, some of its heat-energy is transferable; and the transferable portion may well be regarded as and styled potential.

Available or potential heat-energy readily transfers itself to space or to other bodies, just as compressed air readily leaks out of its reservoir; and in so doing becomes less potential*. Even if it is all received by other bodies in the room, the potential energy of the contents of the room is to the extent of the leakage diminished.

To say that heat-energy is constant in quantity is the same as saying that the molecules of an escaping gas need not part with their energy but may individually retain it. In the abstract either proposition is true, but just as it was only the compression part of gas energy that was potential or available for us, so also it is only the high-temperature part of heat-energy that is similarly available.

It is customary and correct to say that the available part of heat-energy is converted into other forms when allowed to be active and do work. A distinction is thus apparently drawn between heat-engines, on the one hand, and water-engines, compressed-air engines, or electric engines, on the other; because in them the water, the air, and the electricity flow away undiminished in quantity and only at a lower level, pressure, or potential.

* It is tempting thus to use the adjective "potential" in the sense of available. I do not at present wish to justify this secondary usage of the word, but I let it stand as a suggestive and harmless eccentricity.

This distinction is the cause of some confusion or bewilderment, and I verily believe of some incredulity. Prof. Osborne Reynolds has shown, in his interesting biography of Joule, how great were the difficulties felt by the most eminent men in realising that heat actually disappeared in a heat-engine: that less heat was given to condenser than was received from boiler; that not only temperature fell but *also* heat was lost; and we all know how Carnot's theory was based on the contrary hypothesis.

But one mode of avoiding what is reasonably felt to be an artificial and puzzling distinction, is to say that in *all* the engines mentioned transformation of available energy occurs as soon as activity begins, and that that engine is perfect which enables all the energy available to transform itself in the desired way.

The potential energy of the raised water (so to speak for the moment), of the compressed air, of the separated chemicals, disappears, and transforms itself into the motion of a turbine, a bullet, or a motor-dynamo; the potential energy of the hot "working-substance" disappears in like manner, and results in the motion of fly-wheel belts and shafting*.

It is inconvenient to speak of the energy of heat as kinetic. It *is* much of it kinetic, just as the energy of compressed air and everything else may be kinetic; but to us here and now its available portion is not kinetic, but potential. When we can deal with molecules it can be regarded how we please—alternately kinetic and potential, probably, like the energy of a vibrating fork, or on the average half-and-half, like waves;—but till then the kinetic energy of individual atoms is useless, it is the average energy of a group which alone is useful, and this is all that we attempt to utilise in every one of the cases cited. Practically useful kinetic molecular energy exists only when all the molecules are rushing one way, and that is never the case with heat.

By the energy of a spring we mean its energy over and above its useless energy of average temperature; by the energy of a storage-battery we mean the portion corresponding to the reduced lead and the peroxide; by the energy of a waterfall we do not mean to include the warmth of the water above absolute zero; we always mean that portion of its total energy which we aim at utilising, and so speaking we say that a watch, or a dynamo, or a turbine, are efficient machines. By

* Prof. Fitzgerald has pointed out to me that this analogy will work well if I use *entropy* as the analogue of the water in a waterfall, because entropy does really fall in temperature and remain constant in quantity while a perfect heat-engine is working. This suggestion I hope to develop along with other thermodynamic matters in a future paper.

the energy of a hot body we ought to mean that portion of its energy which it has over and above its useless energy of average temperature, the portion which it is willing and able to part with, the only portion we hope to use or aim at using; and so speaking we might call a heat-engine an efficient machine.

We get a notion of low efficiency in the one case, of high efficiency in the others—we confuse ourselves sometimes with statements about the dissipation of energy—all because we perversely *attend* to the energy of average temperature in the one case but not in any of the others. We have done it naturally enough, because in that one case our attention was specially directed to the subject of heat; but it may be a help to realise that all the cases are essentially similar and on the same footing.

True a steam-engine and boiler is *not* an efficient arrangement, only about 8 or 9 per cent. at the best, but that is because of the great unnecessary drop of temperature between furnace and boiler; starting with the temperature of the *boiler*, and ignoring all energy below the temperature of the condenser, it may be efficient enough, 80 or 90 per cent. I suppose. It is better not to pretend to be able to use average molecular energy until we have learnt how to do it.

The portion of heat which can at a single operation be converted into work is very nearly the same as that which leaks away when the body is allowed to cool; just as it is the potential portion which disappears from a standing Leyden jar, or a running-down weight, or a rusting spring, or a leaky reservoir. The difference between the cases is that whereas the capacity of Leyden jars and tanks is constant, capacity for heat is apt to vary with the other conditions of a body; hence intrinsic energy is not solely a function of temperature, but subordinately of volume also. This fact necessitates caution before the above statement can be regarded as complete.

What a body will freely yield throughout a cycle of operations, that can be utilised. We cannot advantageously gain energy by pumping it. Whatever must be pumped is unavailable. We require an artesian well or automatic supply of energy if we are to get work out of it.

A further statement also is necessary if we are to concern ourselves with practical modes of utilising the theoretically available energy; some such statement as the following:—Available energy can only be continually utilised by means of reversible operations. Unless the working-substance is restored to its original condition the process cannot continually go on; and any operation which is not reversible involves dissipation of energy or needless waste of availability.

That the available portion of heat should bear to the whole heat the same ratio which the available drop of temperature bears to the absolute temperature, is essentially but a definition of temperature; it is an assertion that temperature is best measured as proportional to heat, or that the zero of temperature may conveniently be taken to correspond with the zero of heat.

An essential part of the second law of Thermodynamics therefore proceeds to state itself in a very general and purely commonsense form, thus:—

The portion of energy which a body can automatically part with is alone available for doing work; and only that portion which is parted with reversibly is actually utilised.

There is no need to mention "heat;" it is equally true of every form of energy. When a cell has run down, or a reservoir leaked itself empty, as empty as it wants to, any further energy it may have is useless; and any portion which flowed out in an uncontrolled or irreversible manner will have been wasted.

There is something specific to be said about each form of energy in order to apply the above statement definitely to that form; and in the case of heat the supplementary statement needed is that heat will not automatically leave a body for others at higher temperature: if it goes to a hotter body it must be carried by matter, or electricity, or something else, so that it is not a pure and simple flow of heat. In other words, heat will not flow "uphill" by pure conduction; and conduction is the only mode of automatic conveyance of heat as heat*. Water, air, or electricity can flow "uphill" for a time and can do work at the same time, by reason of their property of inertia. Heat cannot: it has no inertia. None of the uphill processes can go on continually or cyclically†.

The law of dissipation of energy states itself thus:—

If a body has any portion of energy in such condition that it is able irreversibly to leave the body, that portion usually does leave, sooner or later. This is only a rewording of the customary statement that the potential energy of a system tends towards a minimum; or, really, except that circumstances often delay the consummation unpractically long, towards zero. The universe will be stagnant, though by no means stationary, when its potential energy is nothing. There

* Radiation is not heat, but another quite distinct form of energy. The phrase "radiant heat" is responsible for immense confusion.

† Work and heat may be coaxed out of a body below the temperature of surrounding objects, as, for instance, by letting air escape from a high-pressure reservoir; but such a process is not cyclical until the air is put back again, and by that operation the heat has to be put back too. Heat by itself cannot flow uphill at all.

will then be abundance of motion but no force, and therefore no "activity."

To make use of the readily detachable portion of energy may not be a very simple thing, and commonly requires a machine, sometimes an ingenious machine. Give a savage a charged Leyden jar, and he will probably detach from it its available energy pretty soon. But give him a charged storage-battery and he will not know what to do. A bit of thin wire, however, is all the mechanism absolutely needed in order to afford him some light and heat; if mechanical motion is required then some form of dynamo or electromagnetic motor must be supplied. Often and often we do not know how fairly to utilise even the portion of energy automatically streaming off from a body—from a gas-flame for instance when we need light. To utilise more of a body's energy than it will part with is impossible; but the progress of science may conceivably teach us, not only how to utilise the whole of what bodies already freely give off, but also perhaps even how to make bodies (say molecules for instance) part with much energy which at present, if left to themselves, they permanently retain.

The first and most general portion of the second law of Thermodynamics, stated above in italics, will always remain true, even when the second part, about the non-uphill flow of heat, has by future discovery been upset; because the application of a machine for the purpose of extracting otherwise retained energy, not by a process analogous to pumping but by enabling it automatically to flow whereas without the machine it could not, can hardly be regarded as other than automatic; else would the present machines for directing the flow of already available heat be liable to a similar objection. Even if the contrivance necessary for extracting molecular energy turn out to be a live thing,—and this Dr. Johnstone Stoney* most suggestively conceives it possible may be the function of some bacteria [a remarkably appropriate demoniacal function for the producers of disease], yet life, too, so far as it falls into the scheme of physics, must be considered as an automatic process, and only the energy which by any device a body can be made automatically to yield without pumping can ever be utilised. But statements about heat not flowing up hill, or about not cooling bodies below surrounding objects, or about $\int \frac{dQ}{T} = 0$, or $dQ = Td\phi$, &c.; these are liable to ultimate modification with the progress of science, since the very terms Heat and Temperature are

* Phil. Mag. April 1893.

undynamical blinkers appropriate to the consideration of particles in the lump.

Appendix 1.—*The Objectivity of Energy and the Question of Gravitation.*

Mr. Heaviside* freely contemplates the flux of energy, but declines to admit its identity, or as he calls it objectivity. And he further doubts my proposition about transformation accompanying transference, because, he says, "convection of energy," *i. e.* simple locomotion of stored energy, "is a true flux." So it is, but it is not what I meant by *transfer*. Locomotion is so absolutely essential to translational kinetic energy that I hardly think it can be desirable ever to speak of mere locomotion as transfer, even although the moving thing be a bent bow or stretched spring. It is, however, a question of convenience, and undeniably convection must enter into a flux equation, for a bullet entering a partitioned-off region of space brings into it energy which was not there before, and, when it leaves, conveys it out again.

My proposition amounts to just this, that whatever energy appears in a bounded region must necessarily have passed through the boundary. This, if true, seems to me to confer upon energy the same kind of identity or continuous existence (or if you please objectivity) as matter possesses.

The ordinary law of conservation does not assert or contemplate continuous existence: it has no objection to seeing energy disappear from existence in one place provided an equal quantity reappears somewhere else—say inside a bounded region. Either it does not attend to or believe in the fact of transfer, or else it is satisfied with a kind of fourth-dimensional out-of-space path.

As to "objectivity" or "reality," there are always metaphysical difficulties about predicating that; and Mr. Heaviside's objection that since motion is relative, energy can hardly be absolute, must be allowed due weight. This is a point on which Professor Newcomb has written in Phil. Mag. February 1889; arguing that it actually limits the generality of the law of conservation. I hope some day to discuss this at more length; meanwhile my belief is that it will be ascertained that motion with respect to the ether is the energetic thing and that other absolute motion is meaningless.

The fact (if fact it be) that energy has a continuous existence, or that if it appears in a closed region it must

* Phil. Trans. 1892, p. 427, "On the Forces, Stresses, and Fluxes of Energy in the Electromagnetic Field," by Oliver Heaviside, F.R.S.

have penetrated the walls, is expressed quite clearly by Mr. Heaviside's equation

$$\text{conv}\, eu = \dot{e},$$

where e is the energy per unit volume, u is its velocity of locomotion, and where *conv.* stands for $S\nabla$, or $d/dx + d/dy + d/dz$.

But Mr. Heaviside is not satisfied with this simple equation of continuity, and proceeds to complicate it by introducing:—

(1) Intrinsic sources; *i.e.* creation or fourth-dimensional apparition of energy; of which the chief example is the gravitation bogey, whose path and nature no man yet knows.

(2) Flux of energy travelling not with matter at a definite speed, but in some other way so that its speed is uncertain. For example of this he instances radiation, but surely that has a definite enough velocity. He might have instanced conduction of heat; but there again, treated merely as a flux of energy, the amount crossing unit area per second is definite enough. Mr. Heaviside would probably agree, but would prefer not to analyse it into two factors e and u; and to this I cannot object.

(3) But to his third category Q, the rate of waste of energy, I am bound to object. The insertion of dissipation of energy as if it were a mysterious disappearance term, is open to the objection suggested above against (1), and also to the objection that it unduly elevates the available portion of energy into being the whole of it.

So long as these various terms are only introduced for practical purposes, *i. e.* to direct attention to what might otherwise get overlooked, they are well enough; but they must not be supposed to represent the reality of things. It is true that the case of gravitation, if it be transmitted instantaneously, as seems not unlikely, is a curious simulacrum of action at a distance; whereby of course energy could be generated *de novo* inside closed boundaries readily enough; but infinite speed of transmission only requires infinite incompressibility in a medium, it does not dispense with a medium; and if a medium of transmission exists, as all analogy and coherence urges if not insists, then gravitation is no exception, and its energy must pass through the walls in order to get inside a boundary, although it may pass through at an infinite pace. It may be better, however, not to assume the pace infinite till proved, but to have a term in the most general energy-equations expressive of the possible propagation of gravitation in time, notwithstanding that its speed is unknown and certainly excessively great.

I may refer to another reply I have made to Mr. Heaviside in 'Nature,' vol xlvii. p. 293 (January 26, 1893).

APPENDIX 2.—*More detailed Discussion of the Transmission of Energy in difficult cases.*

When I say (as I do on pp. 14, 15, and 20) that acting bodies have the same velocity, I of course mean their action to be immediate. If indirect action is contemplated, it is too obvious that a clock-weight has not the same speed as the tip of the second-hand or the hammer of its bell; but at the contact of every cog, two wheels, driver and driven, are moving at the same pace. But now, it may be asked, if all action is contact-action, if all action is direct, how is it possible ever to get a variation in speed? This is a question worth answering.

It is done and only done by means of rotation. The type of all such actions is a rotating wheel propelled by an uncentral force. In such a wheel, regarded as a single rigid body, we have every gradation of speed from a maximum to nothing, and we can make use of or transmit elsewhere what speed we like. This is the essence of levers, and mechanism in general: without rotation the speed of all parts is the same, and therefore the same as the point to which the driving force is applied.

But now, treating the wheel as what it is—an assemblage of particles—how comes it that they can act on each other so as to generate differences of speed? How can a force applied tangentially to the face of a sphere cause part of the opposite hemisphere to move backwards?

If we accept the sphere as a rigid body, nothing is easier than to equate the momentum generated to the impulse of the applied force, and its moment of momentum to the moment of the impulse; but if we treat it as an assemblage of connected particles it is not so easy to tackle the problem. As is well known it did historically give trouble, until it was realised, on the ground of Newton's third law (or D'Alembert's Principle as it was called), that all internal stresses balanced each other, and might therefore be ignored for the purpose of deducing the final result.

There is now no controversy as to final result; the only question is how universal contact-action, with equal velocity between agent and patient, or driver and driven, can account for the ultimate result of all grades of velocity through zero even to minus.

There is no need to take refuge behind any such blinkers as D'Alembert's Principle: an assemblage of connected particles can be directly contemplated. Let one of them receive a blow, it is passed on to the others and the momentum spreads laterally by oblique impacts, the amount of obliquity being limited only by the arrangement of the molecules; and the component of the blow or thrust transmitted in any direction is diminished in accordance with the suitable cosine law. The process is notoriously not an easy one to follow into detail, even in a simple case, partly for lack of data ; and there is some uncertainty as to the disposal of the energy for the case of a blow, though even in that case there is no uncertainty about the momentum ; while for a steady force the body, however essentially elastic, gets rapidly into a practically rigid state, and the molecules then merely act as transmitters of the energy. They are, as it were, connected by massless struts and ties, and along these the energy is transmitted, partitioning itself off into several directions, much as it did in the case of impact, and producing local velocities determined by the arrangement of the particles, that is, by the shape and other circumstances of the body*.

The conception of a rigid body, to which a couple can be applied, and which moves as a whole without dislocation of parts, every portion instantaneously feeling whatever force there is to reach it, simplifies problems enormously ; and it may be said that just as a moving body retains its kinetic energy and carries it through space without transfer or transformation, so a rigid body conveys thrust or potential energy through space, receiving it at one point, delivering it up at other points, and transmitting it instantaneously without transfer or transformation. The thrust of a connecting-rod, the torque of a shaft, the tension of a belt, and the tangential stress of a cog-wheel, are typical instances of this practically instantaneous communication, or locomotion of potential energy, caused by a rigid body.

The simplest way to think of the ordinary case of gearing and shafting is thus to ignore its molecular structure and treat it as a linkage of entirely rigid bodies, where the potential energy communicated to one point is conveyed elsewhere as a simple flux without transfer or transformation, as kinetic is conveyed when a bullet is shot across an empty space.

But that this blindfold treatment does not exhaust the matter can be seen at once by thinking of a moving fly-wheel suddenly geared on to a stationary cog-wheel, so as to transmit a portion of its kinetic energy to machinery. The stress necessary to effect the transfer is too gigantic, and results in damage unless some elasticity is provided in shaft or spokes wherewith to store it temporarily and subsequently give it

* In Thomson and Tait, vol. i. part 1, §§ 311–318, the effect of an impact on a number of elastically connected particles is treated. So it is partly in a review in 'Nature,' vol. xlvii. p. 601 (April 27, 1893)

out again in the kinetic form, or unless by means of slip a good deal of energy is dissipated. But some dissipation is essential anyhow: a certain fraction, equal to the ratio of the moving mass to the whole mass, must assume a vibrational form of some kind.

It is instructive to recollect certain elementary facts, such as that in the impact of two perfectly elastic bodies, one moving, one stationary, all the energy and all the momentum *may* be transmitted complete, but that in this case the resulting motion of the two bodies cannot be the same. Whereas, if by reason of elastic vibrations, or by reason of heat dissipation, it is arranged that two equal bodies shall after impact move together, then only one half the momentum and only one quarter of the energy is transmitted, the remaining half of the energy having been diverted or wasted in heat or vibration.

So it is with any system of shafting or mill-gearing to which motion has to be communicated from a revolving fly-wheel. The conservation of moment of momentum gives part of the circumstances; the refusal of ultimate mutual recoil gives the other part. And the amount of ultimate vibrational dissipation of energy is precisely the same, under these circumstances, for perfectly elastic as for perfectly inelastic bodies. The case is not unlike the opening of a stopcock between a full and empty pair of reservoirs. A certain fraction of the energy is necessarily either dissipated in heat or left as permanent vibration.

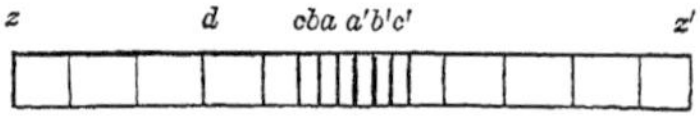

Consider the circumstance of the impact of a couple of equal elastic rods moving end-on. Conceive them in transverse strata labelled $a, b, c, \ldots z$ for each rod. Let one rod be stationary, and the other strike it longitudinally with velocity v. At the beginning of the impact a strikes a', and the two move on in contact with the speed $\frac{1}{2}v$. b now strikes the mass $a\,a'$, and would accelerate it, but at the same instant a' strikes b', communicating the motion to it and neutralizing the acceleration on itself; so that now the four strata are all moving with speed $\frac{1}{2}v$, while all the rest of one rod is still stationary, and all the rest of the other rod is still moving with original velocity v. The length of the half-speed piece about the point of contact continually increases, and behaves as a body under gradually increasing compression as it receives blow after blow on either end. At length z and z' strike, and the recoil is ready to begin. This is the middle of the impact, half the momentum has been transmitted and half the energy; but of the transmitted half energy, one-half again, or a quarter of the original only is in the kinetic form, the other quarter is in the potential or elastic-stress form.

If the rods are inelastic everything so far has happened similarly, but nothing further happens; what I have called the middle of the elastic impact is the end of the inelastic one. Kinetic energy from w and y has been transmitted to w' and y' through the intervention of the quasi-rigid compressed portion of each rod, and energy from z has reached z'. A quarter of the whole energy has been transmitted direct as kinetic, and there is no potential because there is no recoil. One half the original energy has been lost.

If the rods are elastic the recoil brings z to rest and flings z' on with the original velocity v; then y stops and y' flies on after z', and so on, till at the end of the impact the rods part asunder, the first one completely stationary, the second one completely moving with all the energy simply kinetic.

If the rods are partially elastic some of the potential energy is dissipated and some utilised, while if they are of unequal length or material the pulses are not timed similarly in both; the shorter one (supposed the striker) is struck dead as before, the longer one is left with a pulse in it after they have separated, and its residual potential energy then assumes the sound-vibrational form: the strata progressing jerkily for some time.

If the rod to be moved is incompressible, its pulse travels instantaneously and it all gets moved at once. A blow to such a rod transmits energy instantaneously, and all in the kinetic form, but there is nothing in mere speed to affect the amount transmitted.

By the consideration of instances we have thus been led to the induction that energy can be transmitted without obvious change of form by substances with infinite properties, *e. g.* by an incompressible solid; all molecular processes being either non-existent or being ignored; but that with ordinary matter there is always some percentage of obvious transformation, though we may apparently have all grades of it from complete to very small.

Thinking of these impact cases alone, it might appear as if I had been overhasty in saying that the whole of energy must be transformed when it is transferred. Yet observe that it has to pass *through* the intermediate condition. A row of ivory balls in contact has another thrown against one end, and from the other end one leaps off. The energy has been transmitted through the row somewhat as it is transmitted through the compressed strata of two impinging rods. Yet if the elastic connexions of every stratum are attended to, and if these be regarded as massless, I think it will be found that all the transmitted kinetic has really passed through a momentary existence as potential. The fact of necessary transformation is not so obvious when you come to look into some of these special cases; but I would refer once more to the proof given at the beginning of Part IV., which seems to me conclusive as to essential fact.

The difficulty arises because when an elastic body is struck (say a massive molecule with a massless spiral spring connexion) it begins to move a little directly the spring is the least compressed, and is moving half speed when the spring is fully compressed; but I venture to say that on any view of the identity of energy the bit of kinetic which it first attains is a bit of energy that has been transmitted through the elastic stress of the spring, and that just as the second half of the energy must admittedly exist in the spring before it can reach the mass, so the first half has already passed through the spring and has reached the mass only after transmutation, although the transformation is disguised while the transference is obvious.

I have now written enough to emphasize what I want to bring forward as the simple doctrine of energy. Some years ago * I attempted it with brevity, but failed to make it clear or to call proper attention to it. Now I have set it forth at length, with illustrative cases, as matter for discussion. It remains to try and formulate briefly and strictly the extended laws of motion appropriate to our present knowledge.

* Especially in Phil. Mag. for October 1879, page 278 *et seq.*, for January 1881, p. 37, and June 1881, p. 530, and for June 1885, p. 482. Also in 'Elementary Mechanics' (Chambers), which was written in 1876 and revised about 1884, without change in the energy chapter so far as I remember; I am not responsible for dates on title-pages.

My attention has just been called to a "Smith's Prize" essay by Mr. R. F. Muirhead, communicated by Professor James Thomson to the Philosophical Magazine for June 1887. This essay, both in itself and in its numerous quotations and criticisms, is an instructive and useful summary or exposition of nearly everything that is foggy, confusing, and utterly unsatisfactory in the fundamental treatment of Dynamics. It is hardly too much to say broadly that the entire order of ideas in that essay is antipodal to the conceptions I am endeavouring to urge on the acceptance of Physicists.

Made in United States
Troutdale, OR
11/03/2026

56844505R00024